Western Center Community Foundation
2345 Searl Parkway, Building 2
Hemet, CA 92543

Edited by Dr. Brett S. Dooley, Krystle Drouault, Dr. Alton Dooley & Brittney E. Stoneburg

Formatted by Brittney E. Stoneburg

ISBN: 978-1-7363640-2-4

Based on the following scientific article:

Dooley AC Jr, Scott E, Green J, Springer KB, Dooley BS, Smith GJ. 2019. Mammut pacificus sp. nov., a newly recognized species of mastodon from the Pleistocene of western North America. PeerJ 7:e6614 https://doi.org/10.7717/peerj.6614

A FOSSIL JOURNEY:

The Story of the Pacific Mastodon

Written By
Dr. Brett S. Dooley
&
Krystle Drouault

Illustrated By
Charlotte J. H. Hohman

WESTERN SCIENCE CENTER

Introduction to Scientific Papers

When scientists discover something new, they tell other scientists about it by writing a paper in a journal; a journal is a special magazine for papers that are all on the same topic. Journal papers are usually written so that they can only be understood by other experts, so in Fossil Journeys, we are taking those papers and trying to make the discoveries inside them readable for everyone!

About the paper

This book talks about the discoveries that were reported in the scientific paper Mammut pacificus sp. nov., a newly recognized species of mastodon from the Pleistocene of western North America. It can be found at: https://peerj.com/articles/6614/.

Paper Authors

Dr. Alton C. Dooley, Jr.
Alton is the Executive Director of the Western Science Center. Follow him on Twitter: @AltonDooley

Eric Scott
Eric is a vertebrate paleontologist who works for Cogstone Resource Management. Follow him on Twitter: @captainfossil

Dr. Jeremy Green
Jeremy is a vertebrate paleontologist with Kent State University in Ohio.

Kathleen Springer
Kathleen works for the United States Geological Survey. Follow her on Twitter & Instagram: @kathspringer

Dr. Brett S. Dooley
Brett is an educator at the Western Science Center. Follow her on Twitter: @bdooley72

Dr. Gregory Smith
Gregory is an instructor for the Interdisciplinary Science and Research Program at John Overton High School. Follow him on Twitter: @paleodentist

People have been studying fossils in the United States for almost 250 years - that's before the Declaration of Independence was written!

A lot has been learned about the animals that lived during the last Ice Age, which scientists call the Pleistocene Epoch. The Pleistocene is the time from 2.6 million years ago to 10 thousand years ago.

ELEPHANT MAMMOTH MASTODON

You have probably seen mammoths, sloths, or other famous Ice Age animals in movies, but this is the story of mastodons! Mastodons were relatives of modern elephants, just like mammoths. Mammoths were very closely related to today's elephants, but the mastodons were not as close. Mastodons were about as close to an elephant as a dog is to a cat. They were similar, but still very different.

Max was a mastodon who lived during the Ice Age. He was found on the west side of what is now a **reservoir** called the Diamond Valley Lake (DVL).

DVL was made by the Metropolitan Water District of Southern California. Scientists used a process called **carbon-dating** to find the age of the fossils from DVL. The fossils found on the west side of the lake were about 16 thousand years old. The fossils on the east side of the lake were closer to 47 thousand years old.

That sounds really old, but this is millions of years after the time the **non-avian dinosaurs** were alive. Imagine all of that time!

Max is the best-preserved mastodon from the DVL collection. He was an adult male. How can we tell this about a fossil animal? We can tell he was an adult because the ends of his bones were **fused.** This showed he was done growing. We know he was a male from the shape and size of his hips. Also, male mastodons had much larger tusks than females.

When the DVL animals were first collected, scientists thought these mastodons might be smaller than the ones found in the eastern part of the USA, but they had not yet found fossils to prove whether or not they were right.

The director of the Western Science Center (WSC), Dr. Alton Dooley, was working on a talk and he wanted to make a **diagram** showing that Max was the largest mastodon from California, but that he was small compared to eastern mastodons. He looked at the teeth of Max but, to his surprise, Max's teeth seemed pretty small, even though other parts of his skeleton were really large!

To try to understand why Max's teeth looked small, he began to look at all of the mastodon teeth at WSC, then started looking at mastodon teeth from all over California. Then, he started to measure mastodon teeth from throughout the country. This included trips to Louisiana, Indiana, Ohio, Colorado, Wyoming, Oregon, and Washington. You can use a map to find all of these states! This was not the only data used. The director also used information from other scientists.

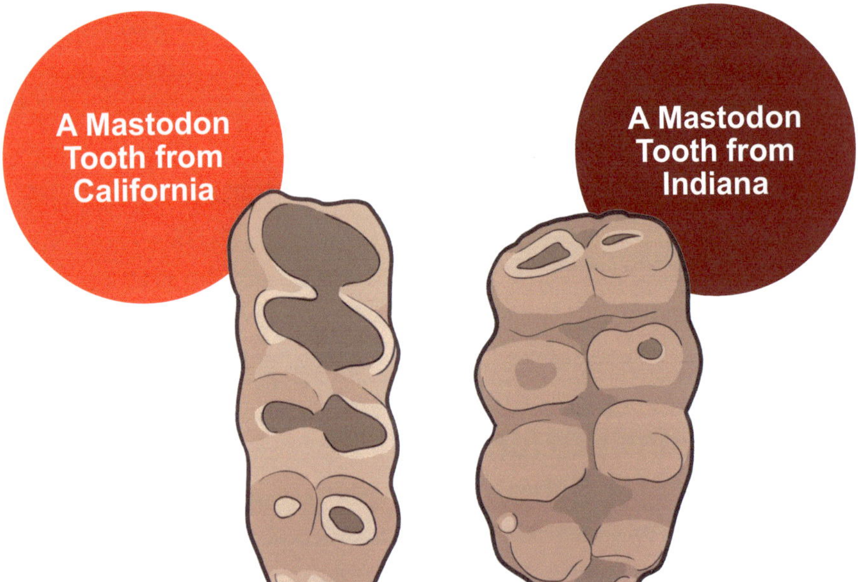

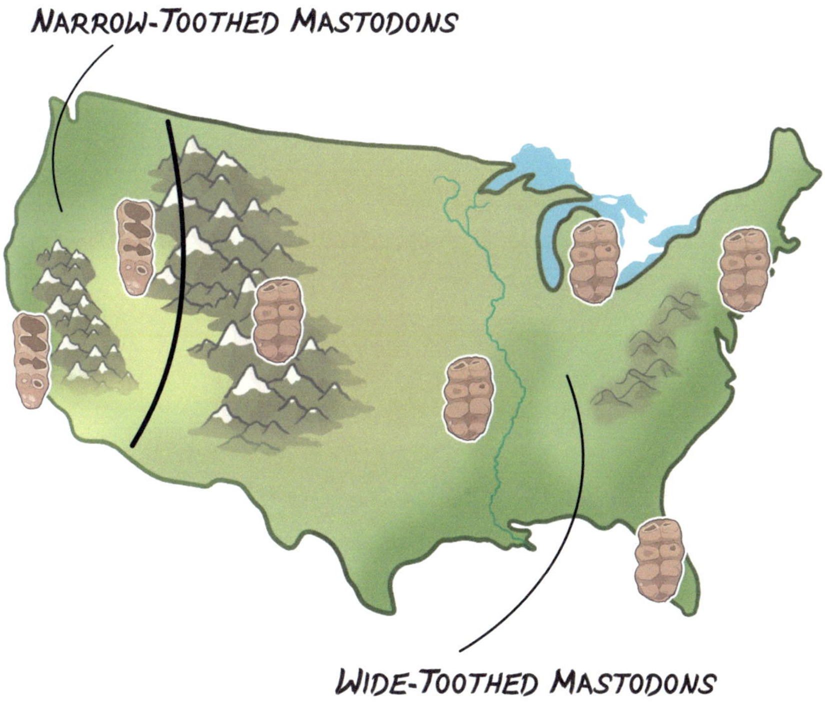

The data about mastodon teeth were taken using **calipers** to measure the maximum length and width of each tooth. What this research showed was that the California and Idaho mastodons had long, narrow **molars** while mastodons from the rest of the country had block-like, wide molars.

The scientists working on the project came up with a number of **hypotheses** to try to explain why the western mastodons were different from the eastern mastodons. The first hypothesis was that the western mastodons had differently shaped molars because they were eating a diet different from the eastern mastodons. This **hypothesis** was later **rejected** because there were as many different types of plants in California as there were where the eastern mastodons lived. Diet could not have been the cause of the molar differences.

Another hypothesis was that the western mastodons were an **isolated population**, separated from other mastodons by large mountains. As a result, scientists thought that these mastodons only had a small range of molar sizes. This hypothesis was **rejected** when it turned out there was a bigger spread of sizes in western mastodon molars than in all of the eastern mastodon molars. Also, narrow toothed mastodons were found on both sides of the mountains.

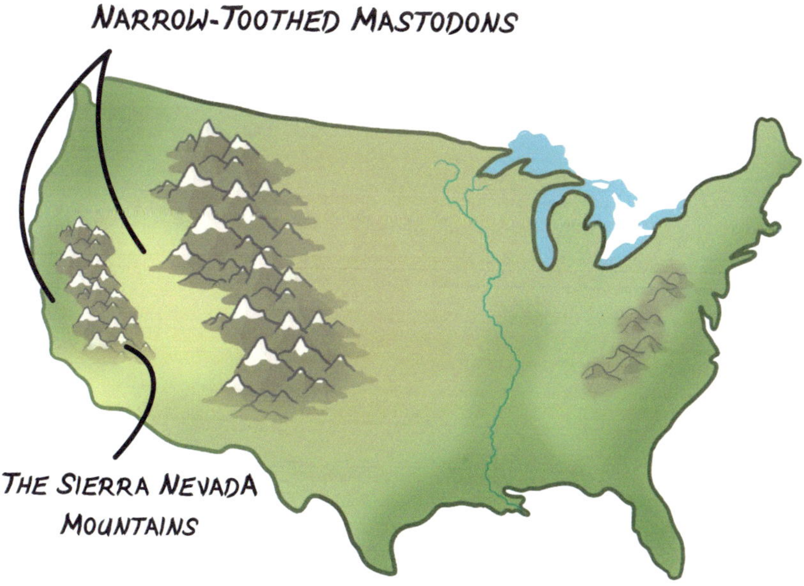

A third hypothesis was that eastern and western mastodons were different because they were actually different **species**. When describing a new species, it is important to explain what **traits** are found in that species. The scientists had already discovered that western mastodons had more narrow teeth than eastern mastodons. To test this hypothesis, the scientists had to see if there were more differences between eastern and western mastodons than just the teeth - and they found a lot!

Besides the narrow teeth, western mastodons also had different hips.

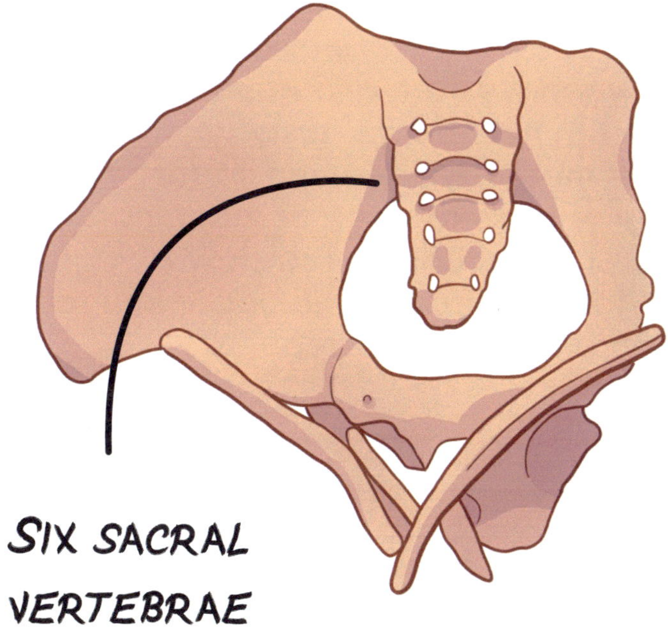

The part of the backbone connected with the hips are the **sacral vertebrae.** There are six of these bones in every western mastodon, but the eastern mastodon usually has five.

Another trait that is **unique** to the western mastodon is that they never have lower tusks. Lower tusks are found in about a quarter of eastern mastodons. The upper tusks' **growth rate** is also different between western and eastern mastodons. Western mastodons' tusks grew more slowly - this means western mastodons had smaller tusks for their age compared with eastern mastodons. The slow tusk growth led to the shape of the upper jaw to be different in the two species.

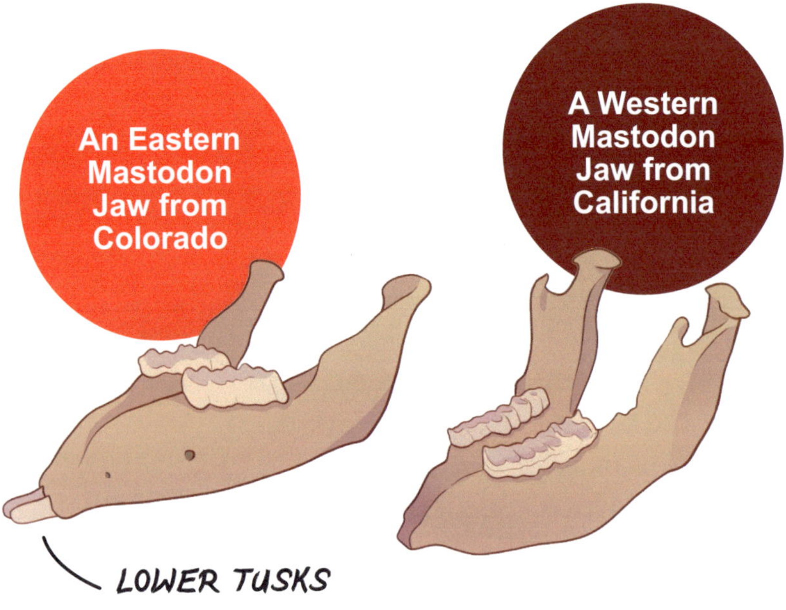

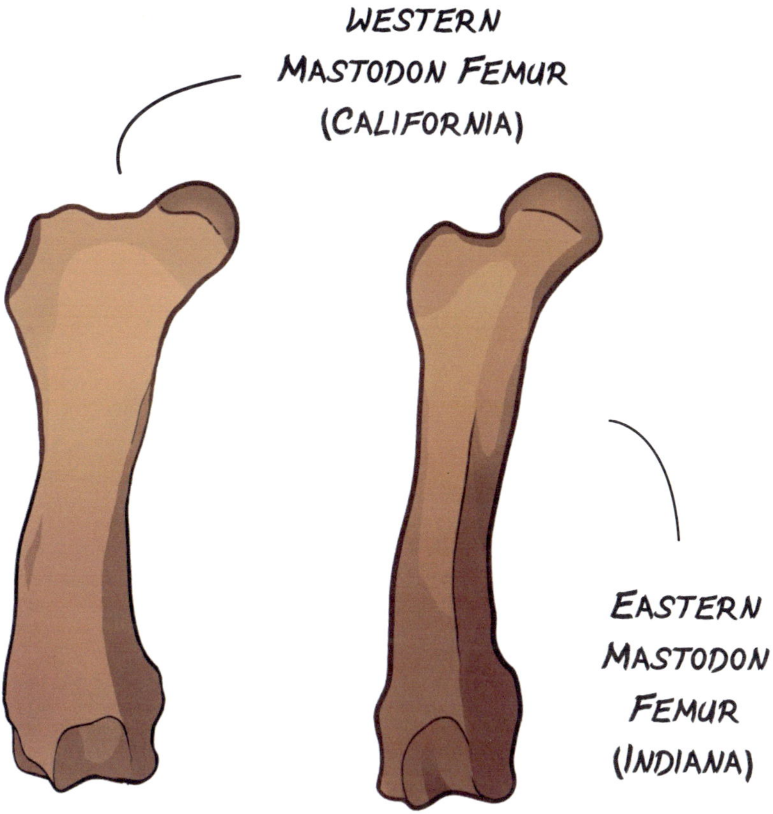

Finally, the **femur** is thicker in the middle in western mastodons than in eastern mastodons. This means western mastodons may have been a little shorter and heavier than eastern mastodons.

Finding all these differences led the scientists to accept the third hypothesis, that the two populations of mastodons are actually different species! Those from the east represent the known species *Mammut americanum*, the American mastodon. Those in the west represent the new species, *Mammut pacificus*, the Pacific mastodon. Fossils from museums across California were used to determine this new species. This is why museum collections are so important!

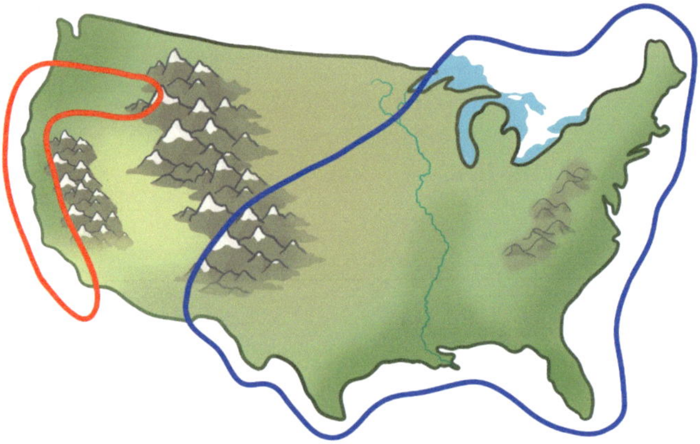

PACIFIC MASTODON RANGE

AMERICAN MASTODON RANGE

With all of these physical features, what exactly did the Pacific mastodon look like? This is hard to say! There are no complete skeletons known. For example, no Pacific mastodon fossil ever found has all of its foot and hand bones preserved. However, we do know from the fossils found that males were about 30% larger than females. We also know that females had smaller, straighter tusks than the males.

MALE MASTODON FEMALE MASTODON

Thanks to the hard work of scientists, the Pacific mastodon is the first new mastodon species named in North America in 50 years! Even after all of this time, there are still unanswered questions to ask about these wonderful animals. Scientists will continue to study Pacific mastodons for many years to come. Maybe someday you will study Pacific mastodons, too!

If you would like to visit **Max the Mastodon** yourself one day, go to the **Western Science Center** in **Hemet, CA,** the natural history museum where Max is now displayed!

Glossary

Calipers - a special ruler-like tool

Carbon dating - a process that uses the ratio of Carbon-14 to Nitrogen-14 atoms to find the age of a sample

Diagram - a two-dimensional shape that explains an object or process

Femur - thigh bone

Fused - joined together physically

Growth rate - how quickly your body gets bigger

Hypotheses - educated guesses; plural of **hypothesis**

Isolated - left alone; away from others of the same species

Molar - teeth in the back of the mouth used for grinding food

Non-avian dinosaurs - any dinosaur you can think of except birds (which are dinosaurs)

Populations - a group of animals living together

Reject - throw away; discard

Reservoir - a body of water that is natural or human-made that forms due to the presence of a dam

Sacrum - the hip portion of the body; **sacral** - having to do with the sacrum

Species - the largest group of organisms that are capable of having fertile young (example - house cats are a species different from cheetahs, leopards, tigers, lions, or other felids.)

Traits - features; characteristics

Unique - being the only one of its kind

Vertebrae - bones of the spine

ABOUT THE AUTHORS

Dr. Brett S. Dooley

Brett has taught all grades from 3rd through high school, undergraduate, and graduate courses. She is an educator at the Western Science Center, creating lessons, designing camps, preparing specimens, and doing research. She earned her doctoral degree in Curriculum and Instruction in 2020 from Texas Tech University. Brett was an author of the original paper and is a co-author of this book. Follow her on Twitter: @bdooley72

Krystle Drouault

Krystle has spent over 15 years in education, with more than 12 years at the Western Science Center as an Education Specialist. She coordinates the Field Trip and Volunteer program and teaches summer camps, and assists with various museum programs. She is a co-author of this book.

ABOUT THE ILLUSTRATOR

Charlotte J. H. Hohman

Charlotte is a Montana State University undergraduate student majoring in geology with a focus in paleontology. She has always loved drawing and has been drawing prehistoric animals for two years. Follow her on Instagram: @charlottejhh

Made in the USA
Las Vegas, NV
07 November 2026

57672507R00017